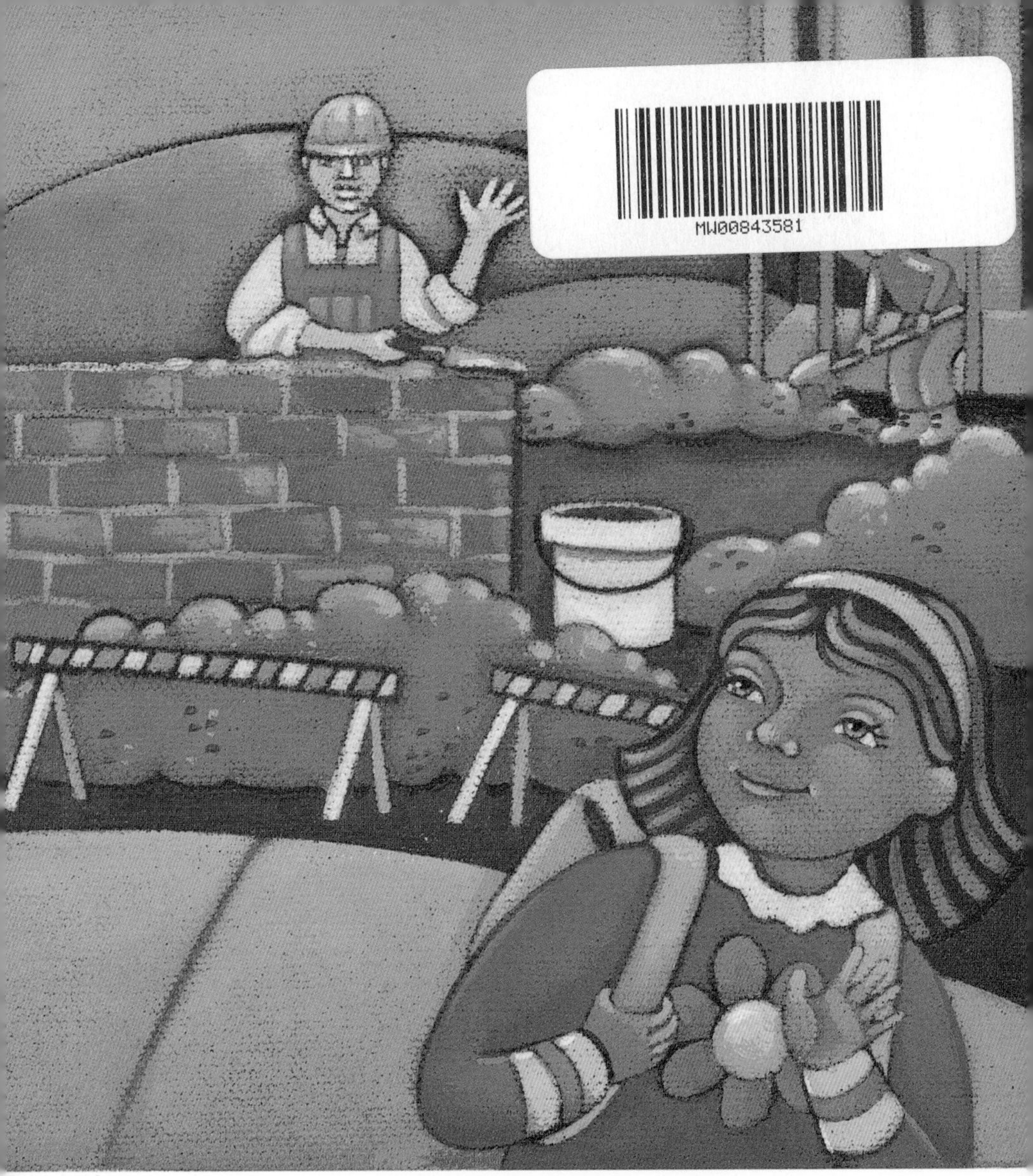

A man sets red bricks.

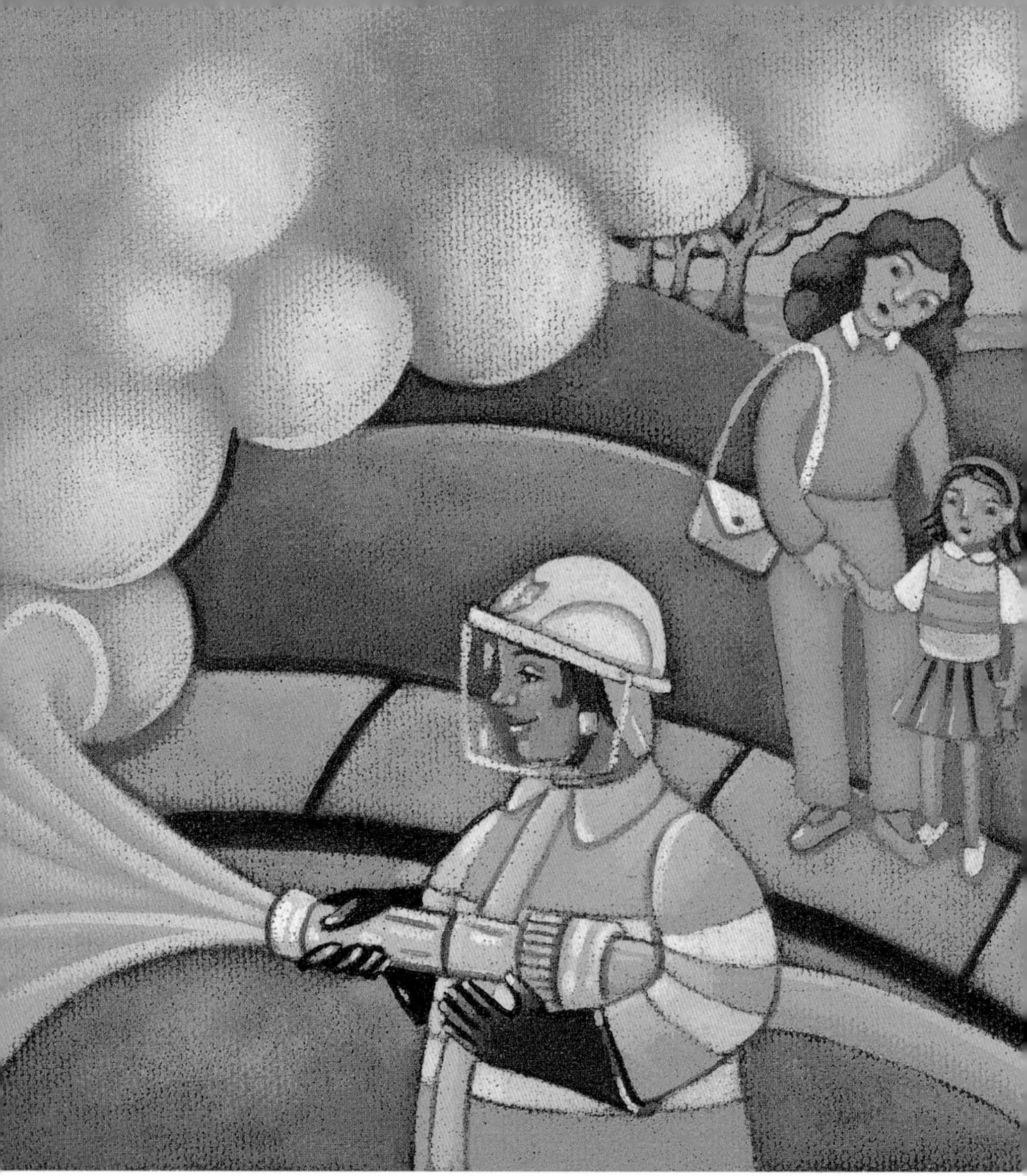

"What do I smell?" I ask.

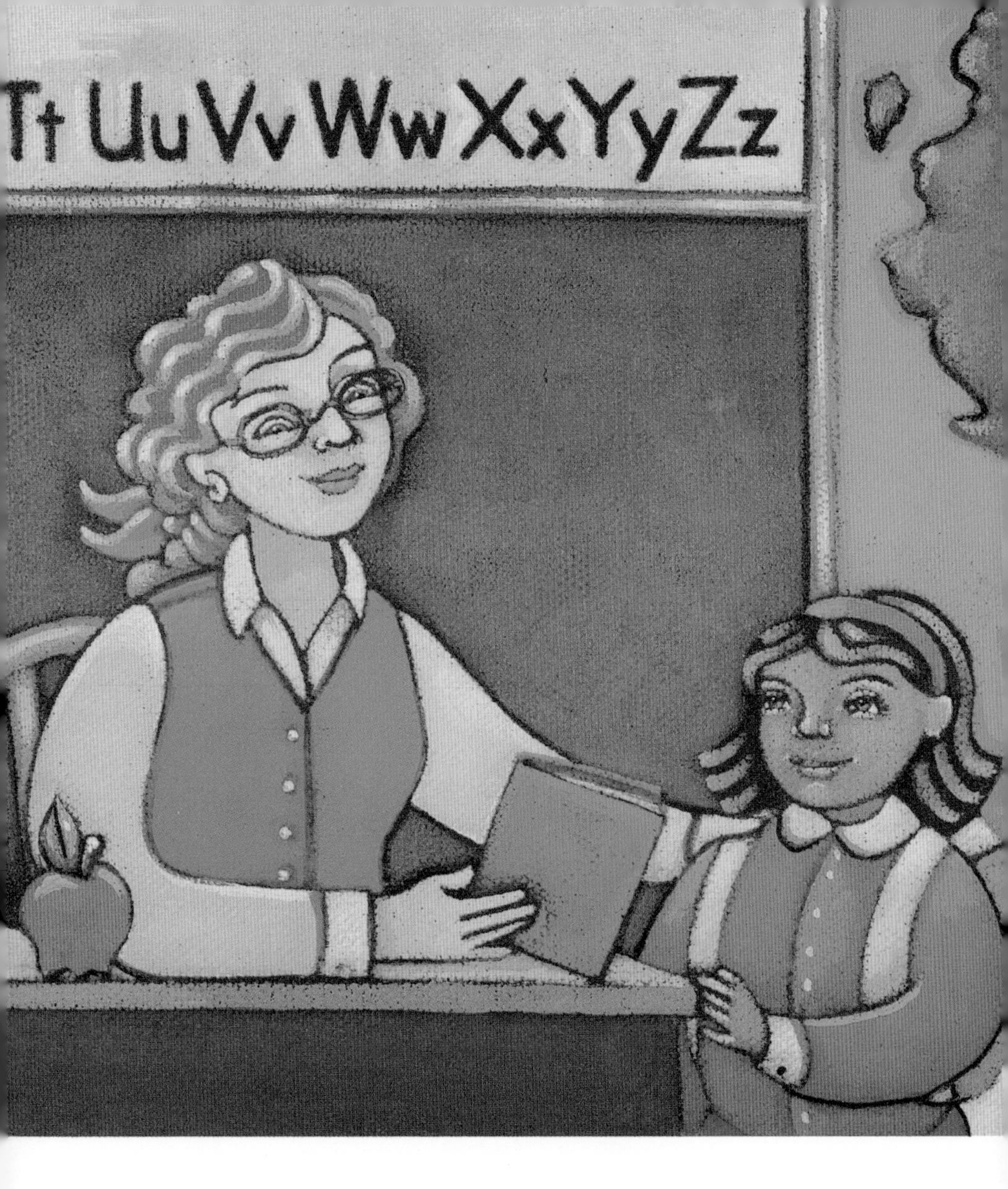

"I help you spell."

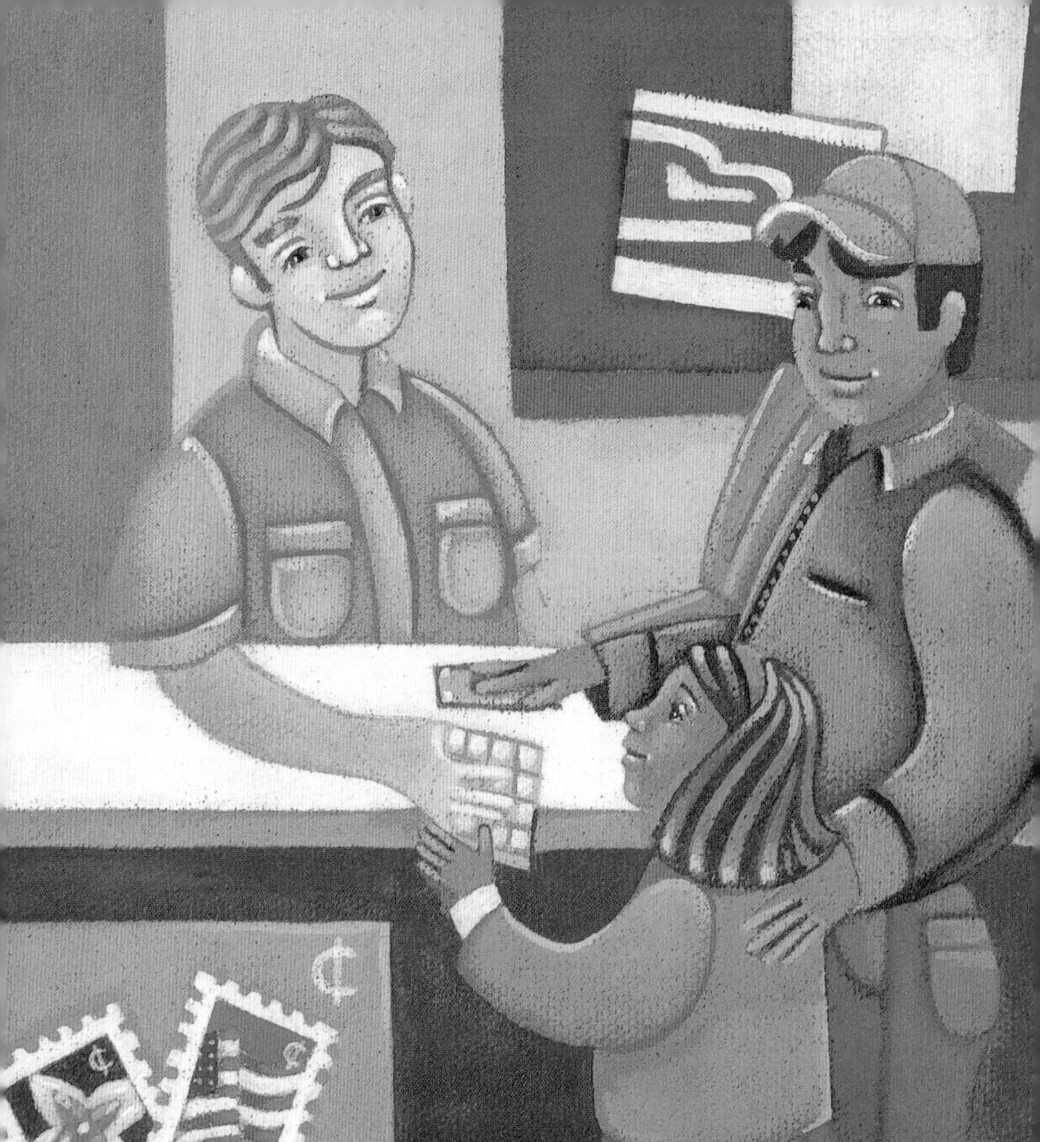

"Get ten stamps from him."

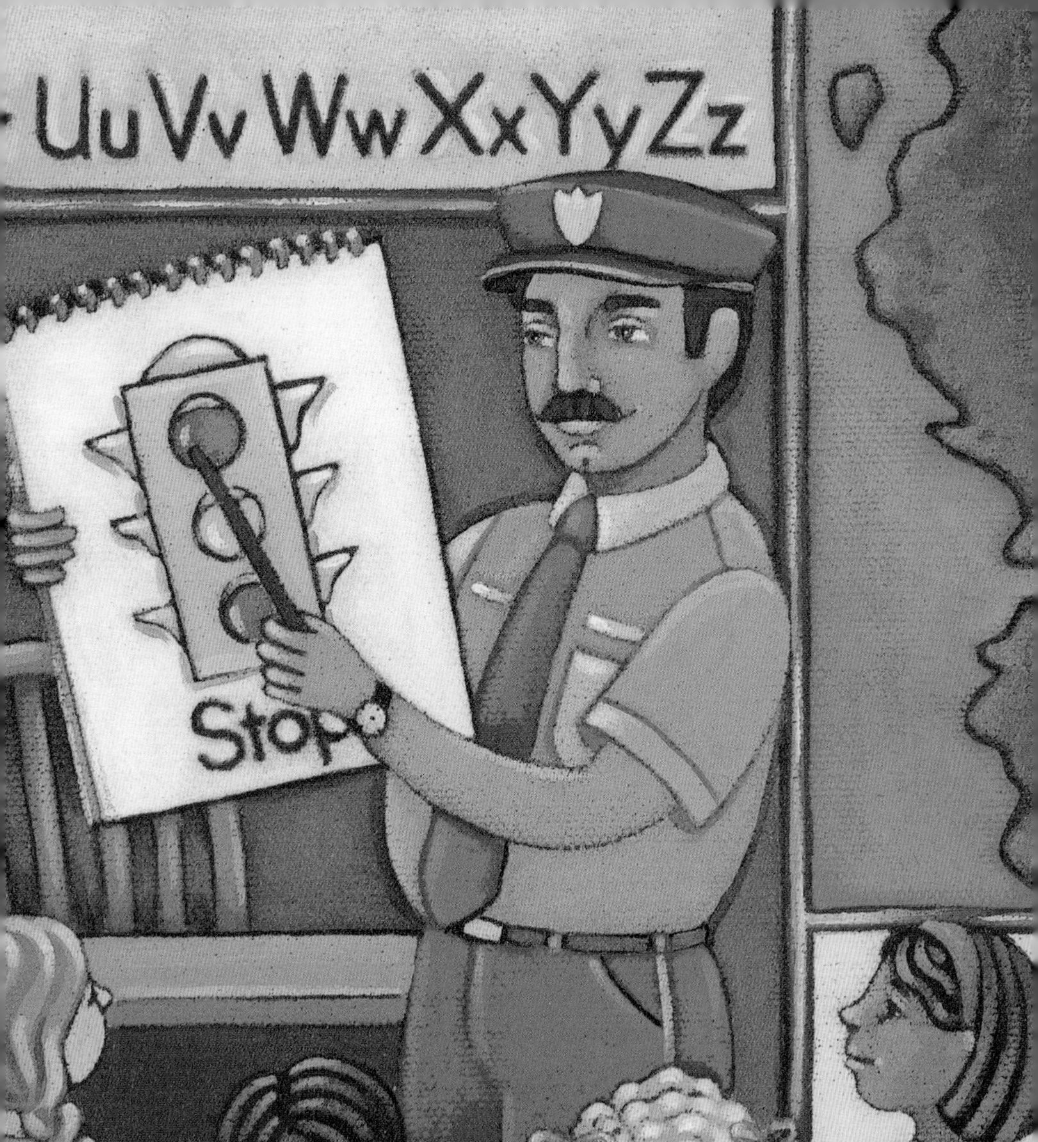

He said, “Stop on red.”

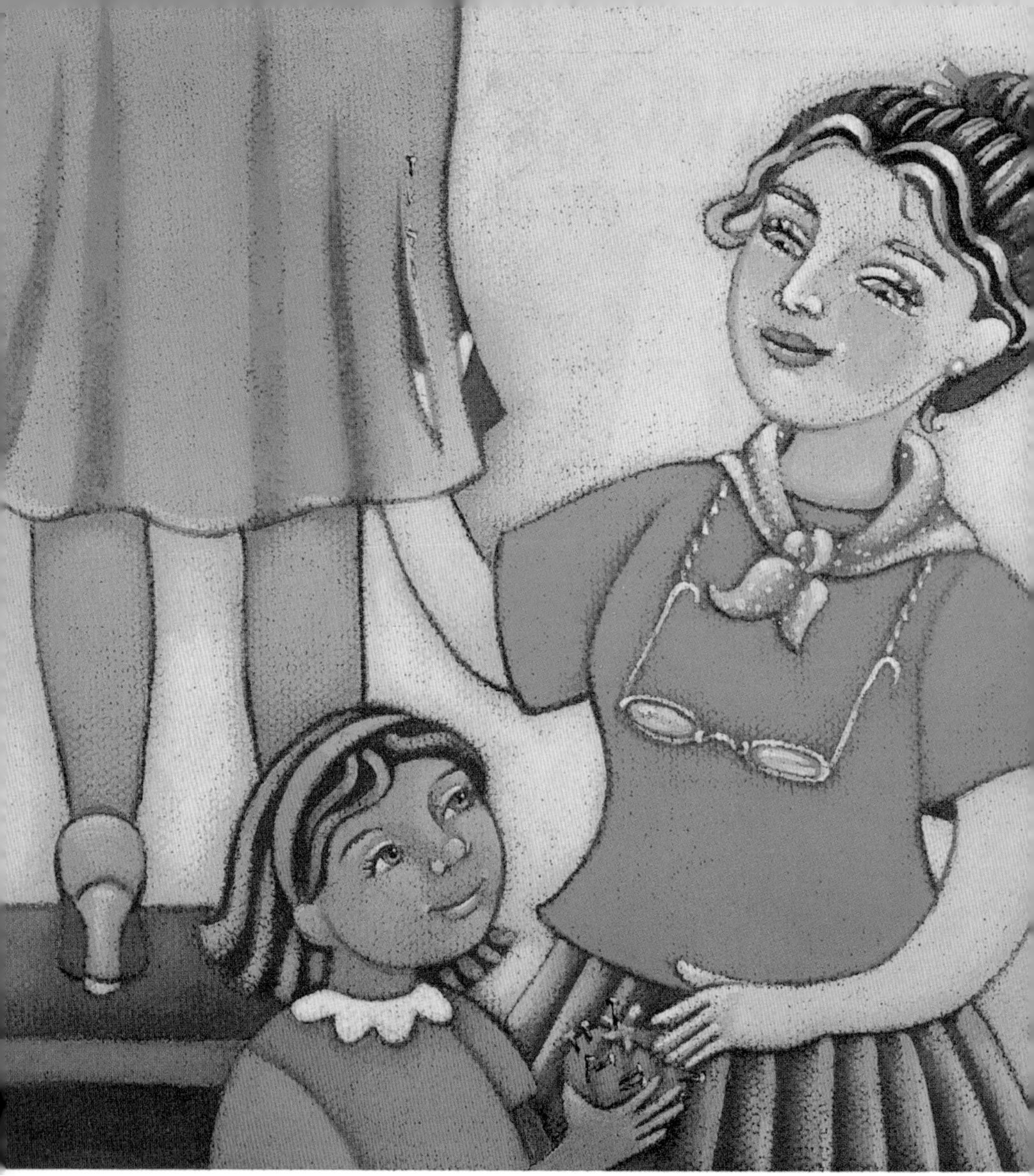

"Help me pin the dress."

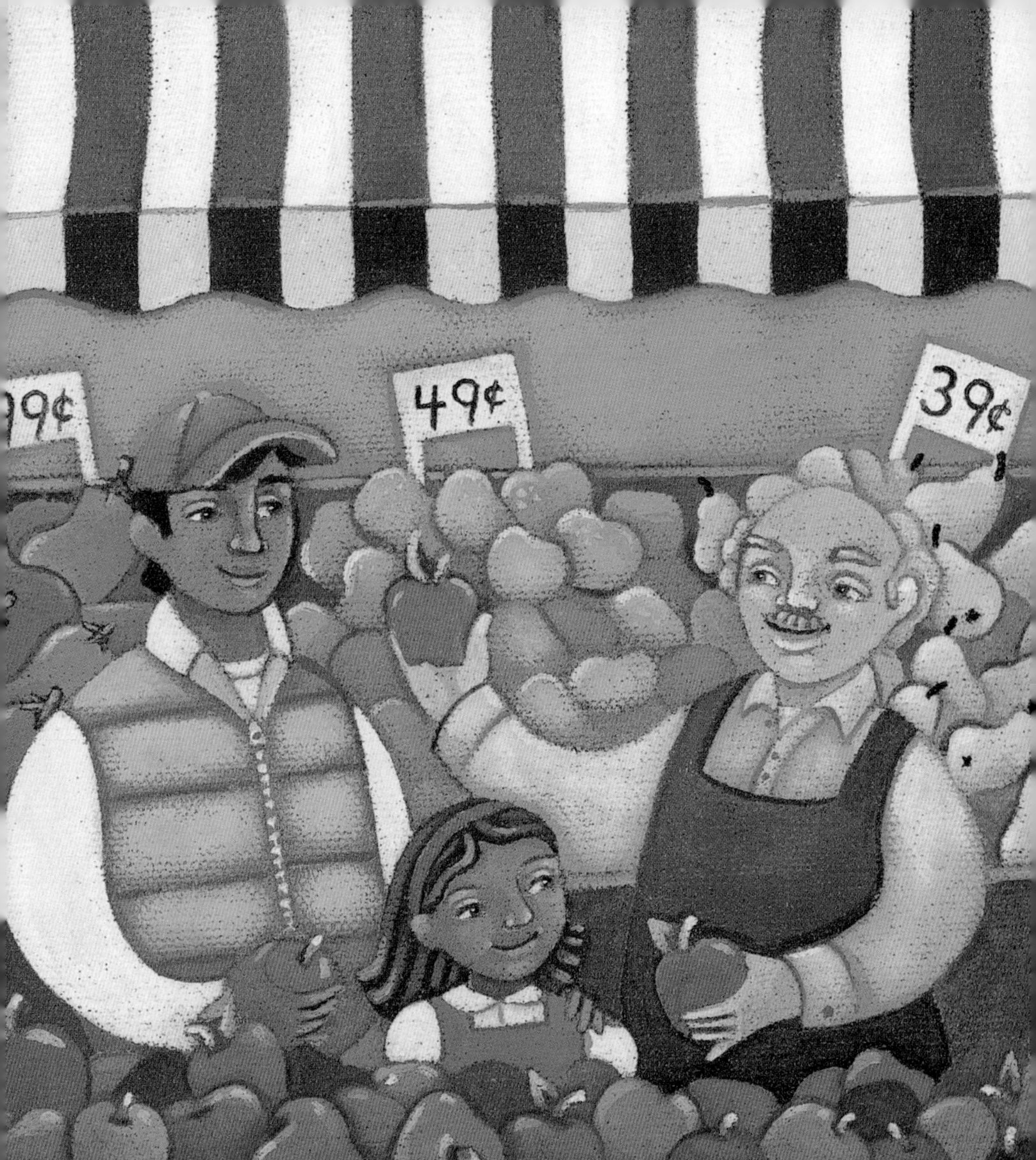

"I can pick the best."

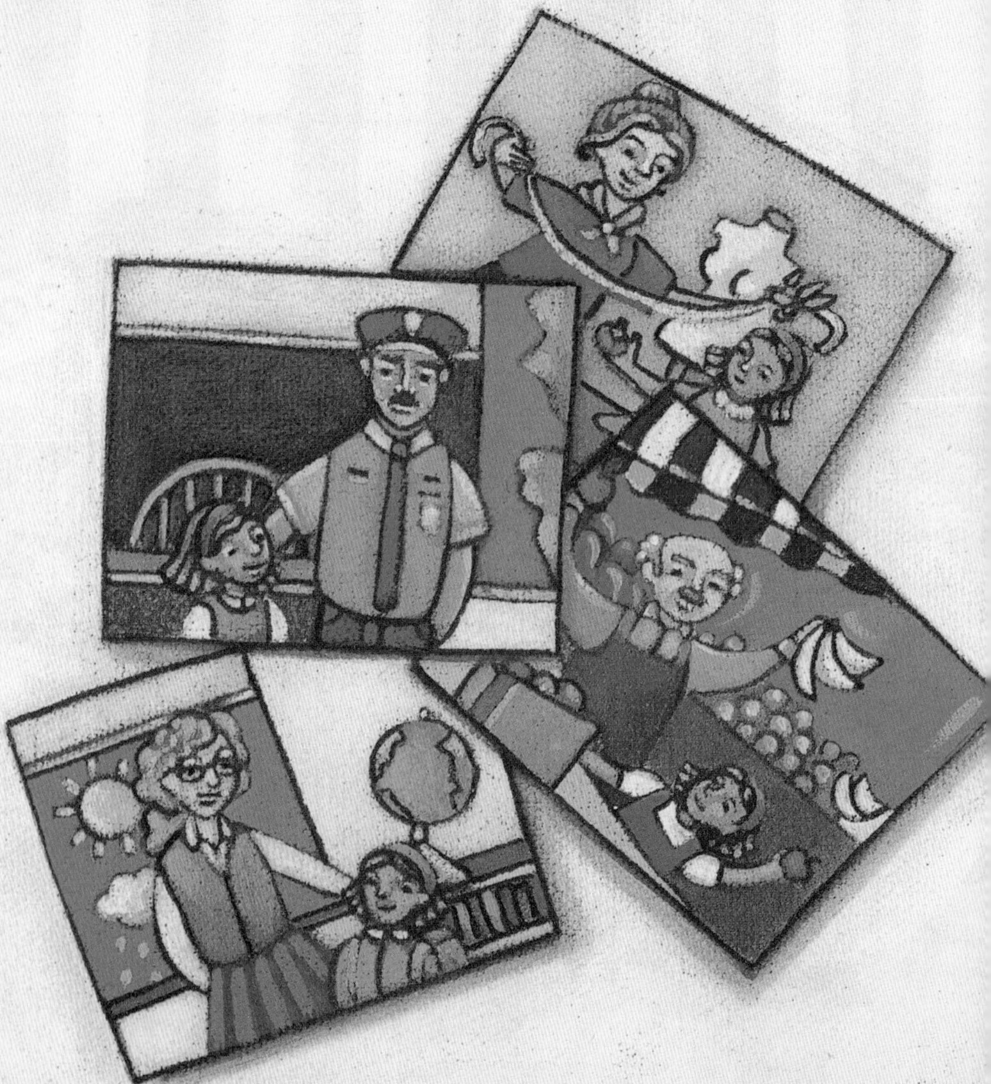

Friends help.